Campden & Chorleywood Food
Research Association Group

Key Topics in Food
Science and Technology - No. 1

Molecular methods in food analysis
- principles and examples -

Leighton Jones

Campden & Chorleywood Food Research Association Group comprises
Campden & Chorleywood Food Research Association
and its subsidiary companies
CCFRA Technology Ltd   CCFRA Group Services Ltd   Campden & Chorleywood Magyarország

Campden & Chorleywood Food
Research Association Group

Chipping Campden, Gloucestershire, GL55 6LD UK
Tel: +44 (0) 1386 842000   Fax: +44 (0) 1386 842100
www.campden.co.uk

© CCFRA 2000
ISBN: 0 905942 28 0

SERIES PREFACE

Food and food production have never had a higher profile, with food-related issues featuring in newspapers or on TV and radio almost every day. At the same time, educational opportunities related to food have never been greater. Food technology is taught in schools, as a subject in its own right, and there is a plethora of food-related courses in colleges and universities - from food science and technology (FST) through nutrition and dietetics to catering and hospitality management.

Despite this attention, there is widespread misunderstanding of food - about what it is, about where it comes from, about how it is produced, and about its role in our lives. One reason for this, perhaps, is that the consumer has become distanced from the food production system as it has become much more sophisticated in response to the developing market for choice and convenience. Whilst other initiatives are addressing the issue of consumer awareness, feedback from the food industry itself and from the educational sector has highlighted the need for short focused overviews of specific aspects of food science and technology with an emphasis on industrial relevance.

The *Key Topics in Food Science and Technology* series of short books therefore sets out to describe some fundamentals of foods, food analysis and food production. In addressing a specific topic, each issue emphasises the principles and illustrates their application through industrial examples. Although aimed primarily at food industry recruits and trainees, the series will also be of interest to those interested in a career in the food industry, FST students, food technology teachers, trainee enforcement officers, and established personnel within industry seeking a broad overview of particular topics.

PREFACE TO THIS VOLUME

Within 50 years of the discovery of the structure of DNA, biotechnology has become a huge international industry. It is delivering great benefits but posing equally significant dilemmas. Molecular diagnostics provides a good example of this. On the one hand it is contributing enormously to clinical diagnosis of disease, forensic analysis of crime scenes and criminals, and analysis of foods. On the other hand, it can deliver diagnoses of incurable inherited diseases, raising life-changing dilemmas for the individuals involved and wider-spread anxieties about potential misuse of personal information by government, employers and insurance companies, for example.

In the context of food analysis, biotechnology is delivering many innovations. Keeping pace with these is not easy, not least because of the proliferation of jargon. This short book sets out to explain the general principles of biotechnology-based diagnostic tests and to illustrate these with examples of how the tests are applied in food analysis. It is hoped that by gaining a clearer understanding of how these tests work, industrialists will be better placed to commission analyses, and question the approaches used and the results obtained. The book does not provide a catalogue of off-the-shelf diagnostic kits nor is it intended as a comprehensive review of which methods have been applied to which problems, though the references and suggested further reading will lead to this information which has been published elsewhere.

ACKNOWLEDGEMENTS

In compiling this guide I am grateful to my colleagues Steve Garrett, Dr. John Holah, Dr. Roy Betts, Beth Hogben, David Dawson, Jon Bird and Dr. John Dooley for their constructive comments and advice. Particular thanks are due to Janette Stewart for the artwork and design.

CONTENTS

1. INTRODUCTION

Molecular methods have progressed from highly technical laboratory procedures, used only as research tools, to simple, reliable and robust one-step procedures that can be used without any laboratory equipment. One of the clearest examples of how simple and robust some tests have become is provided by the range of over-the-counter diagnostic kits designed for use in the home - such as pregnancy test kits.

Since the discovery of the structure of DNA (less than fifty years ago) our understanding of molecular biology has exploded - hand-in-hand with the development of new methods for probing the molecular workings of animals, plants and microbes. These advances first brought benefits in the clinical and forensic fields, but more recently have been widely exploited in the analysis of food and food contaminants. Many diagnostic systems, based on antibody and DNA techniques, are available commercially, and many, many more have been demonstrated and published; new applications based on improved reagents or better engineering are published almost daily.

The purpose of this guide is to explain the basic principles of how these tests work - but it is not a catalogue of the latest tests available, as such publications already exist. It focuses on three major technical areas which dominate developments:

- ◆   Antibody methods (immunoassays)
- ◆   DNA methods
- ◆   ATP methods

In each case, the section begins by describing the basic ingredients of the methods and then outlines how these are combined in different ways to create useful tests. These descriptions are supplemented with some examples of commercially available tests or services based on these principles. Several case studies are also included as 'boxed essays' to illustrate the industrial application of some of the tests.

2. ANTIBODY METHODS

Immunoassays, or antibody-based tests, form the basis of a wide range of simple tests for the rapid and reliable detection of individual components in complex food mixtures. Applications include tests for rapid detection and/or identification of the most important food poisoning micro-organisms that can contaminate foods, determining the species of origin of meat, checking the gluten-free status of food for coeliacs, and identifying natural toxicants or chemical contaminants in foods. In understanding immunoassays it is useful to first look at the ingredients of the tests, and then the ways in which these are assembled to create different test formats.

2.1 Immunoassays - the main ingredients

Antibody-based tests (or assays) are called immunoassays because the one thing that they all have in common is that they use antibodies - proteins produced by the immune system of animals. In addition to antibodies, most immunoassays also have two other vital components: a signal system (e.g. a colour change) to make the test visible, and a solid phase (e.g. a test well, filter paper or dipstick) on which the test takes place. Depending on the test format, the colour change reveals the presence or absence of the target analyte (the component being measured) (see Section 2.2).

Antibodies and antigens

All immunoassays exploit the highly specific interaction between antibodies and antigens (Figure 1). Antibodies are produced by higher animals invaded by a foreign molecule or micro-organism (which, to the immune system, is a complex collection of foreign molecules). The foreign molecule or micro-organism is called the antigen.

Each antibody interacts specifically with the antigen that elicited its production. The two fit together rather like a 'lock and key' (Figure 1). Importantly, however, antibodies also stick tenaciously to their antigens, like glue, to form a bound 'complex'.

Figure 1 - Antibody-antigen interactions are specific

Antibody-antigen interactions are specific: antibodies recognise and stick to their antigen. Although the figure shows one antibody molecule binding to one antigen molecule, in reality the antibody can bind two molecules of the same antigen. This is important for some of the assay formats described later.

During its life, an animal will produce many millions of different antibodies in response to an equally bewildering array of antigens. In everyday terms, this specificity in the antibody-antigen interaction is familiar to anybody who finds themselves immune to a disease caused by one micro-organism (e.g. measles virus) but who fall ill due to infection with a different microbe (e.g. chicken pox virus). The 'glue element' is also an important part of the immune defence, as it enables the antibodies to trap the antigen, for mopping-up by other components of the immune system.

Signal systems

The second main component in an immunoassay is the signal system; this tells the user when the assay is complete and also whether or not the antigen is present. The most common signal system in use is a colour change brought about by an enzyme (Figure 2), but there are many others as described later.

Figure 2 - Enzyme-driven colour change

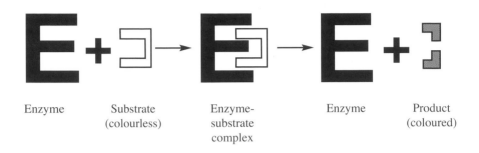

| Enzyme | Substrate (colourless) | Enzyme-substrate complex | Enzyme | Product (coloured) |

The enzyme and substrate (colourless) lock together to provide ideal conditions for conversion of the substrate to a product (coloured). The product is released from the enzyme, freeing the enzyme to catalyse the conversion of remaining substrate to product.

Table 1 - Typical enzyme-substrate combinations used in immunoassays

Enzyme	Substrate	Product
Alkaline phosphatase	para-nitrophenyl phosphate (PNPP)	Yellow solution
	Bromochloroindolyl phosphate / nitro blue tetrazolium (BCIP/NBT)	Black-purple precipitate
Horseradish peroxidase	Tetramethylbenzidine (TMB)	Yellow solution
	Azinodiethylbenzthiazoline sulphonic acid (ABTS)	Green solution
	Chloronaphthol	Blue-black precipitate
	Diaminobenzidine (DAB)	Brown precipitate
	Aminoethylcarbazole (AEC)	Red precipitate

The enzyme catalyses the conversion of a colourless substrate to a coloured product (Table 1). The development of colour is assessed to obtain the test result; the precise meaning of the colour development depends on the test format (see Section 2.2).

A variation on this theme is the use of an enzyme to generate light: the principle is the same in that the enzyme catalyses a reaction in which a substrate is converted to a product, but in this case the reaction releases light energy - that is, a 'glow'. In practice, this system is more widely used in ATP-based hygiene tests, as described later.

There are various alternatives to enzyme mediated colour changes or light emissions. For example, coloured latex beads are often used, as is fluorescence, and sometimes it is possible to see the antibody-antigen 'complex' itself, as a visible white precipitate. To a large extent, the nature of the signal is determined by the third ingredient - the solid phase.

Solid phases

All these reactions - the antibody-antigen binding and the signal generation - have to take place under controlled conditions. This is quite often done in test wells (usually arranged as strips or 'plates'), on dipsticks, on paper membranes, on beads or even in agar gel. Again, the solid phase dictates how the assay as a whole will work, how the signal will be assessed and what the test result actually means. This is all described and illustrated in the later sections.

Wash steps

In most cases, the solid phase does more than provide a physical support for the test - it also allows unbound (i.e. left-over) antibody and antigen to be washed away (separated) from the antibody-antigen bound complex. Similarly it allows other components of the food sample being tested to be washed away, so that they do not interfere with test reaction.

2.2 Immunoassays - typical formats

The sandwich ELISA

The simplest immunoassay format to appreciate is the sandwich ELISA - Enzyme Linked ImmunoSorbent Assay. This is depicted in Figure 3. The term 'enzyme linked' indicates that the antibody is detected through an enzyme-mediated reaction - usually a colour change. 'ImmunoSorbent' means that the test involves antibodies adsorbed on a surface - usually on the surface of a test-well. The term 'sandwich' refers to the fact that the assay involves two antibodies which sandwich the antigen - the first 'captures' the antigen while the second 'detects' the antigen (Figure 3). The end result of a typical sandwich ELISA is shown in Figure 4.

The second antibody (detection antibody), with its enzyme, becomes attached to the well only if the antigen is present to form the sandwich. The presence of enzyme is then detected by adding a colourless substrate which the enzyme converts to a coloured product (Figure 3). The steps involved in a sandwich ELISA are illustrated in Box 1.

Tables 2 and 3 list examples of the kinds of food components and contaminants for which there are commercially available sandwich ELISAs. These assays typically take around 1-3 hours to perform. Note that some of the assays are qualitative (simply indicating whether the antigen is present above a certain threshold or not) whereas other are quantitative (i.e. they can be calibrated to allow quantification of the antigen).

Table 2 - Examples of microbiological food contaminants for which there are commercially available sandwich ELISA test kits

- *Bacillus cereus* diarrhoeal toxin
- *Campylobacter*
- *E. coli* O157
- *Listeria*
- *Salmonella*
- *Staphylococcus aureus* enterotoxin

Common variations on the sandwich ELISA

Although depicted here as the development of a soluble (liquid) product in a test well, assays based on exactly the same principle can work on filter-paper type membranes; in these cases, however, the coloured product that develops is a solid precipitate which remains on the membrane as a dark stain.

An additional variation, which can be used in the test well format or in the membrane based-assays, and which is often used, is the 'indirect' sandwich ELISA. The assay described above involves an enzyme coupled directly to the detection antibody, and is consequently called a 'direct' sandwich ELISA. In contrast, in the indirect sandwich ELISA, the detection antibody does not carry an enzyme. Instead, it may carry a small marker molecule such as biotin. Biotin is easily detected by the protein avidin, which again sticks to it tenaciously. Avidin coupled to an enzyme can therefore be used to detect a biotin-labelled antibody - it just adds another layer to the sandwich (Figure 5).

Figure 3 - The sandwich ELISA

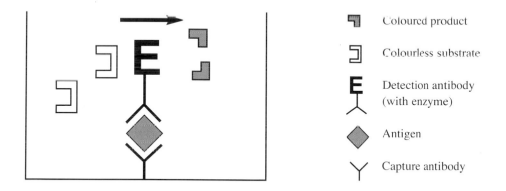

Capture antibody on the test well will capture antigen (if present), which in turn will capture the detection antibody (when added). The detection antibody carries an enzyme which catalyses a colour change. If antigen is present, a sandwich forms and colour develops (see Box 1). If antigen is absent, no sandwich forms and no colour develops. The assay can be quantitative with calibration of antigen concentration to colour intensity.

Table 3 - Examples of food components and chemical contaminants for which there are commercially available sandwich ELISA test kits

Analyte	Notes
Meat species tests - raw	Includes tests for beef, pork, poultry, sheep, horse, rabbit and kangaroo
Meat species tests - cooked	Includes tests for beef, pork, poultry, and sheep
Milk / cheese species testing	For bovine milk in goat or sheep milk
Peanut protein	As a marker for allergenic peanut material
Gluten	Checking gluten-free status of foods for coeliacs

Box 1 - The main steps in a sandwich ELISA

All ELISAs are performed as a series of steps. One of the most crucial steps is washing the test wells between the other steps, to remove unwanted sample and/or reagent. In the sandwich ELISA, the first step is usually addition of sample extract to the well. The well is then incubated to allow the capture antibody to trap its antigen (if present). The well is then emptied and washed (i.e. repeatedly filled and emptied with a mild detergent) to remove the rest of the sample. The detection antibody is then added, and the well incubated to allow the antibody to lock-on to the antigen (if present). Again the well is washed repeatedly to remove unbound detection antibody (and its associated enzyme). Finally, substrate is added - colour only develops if the sandwich has formed.

This approach to the controlled addition and removal of reagents is the basis of most ELISAs (not just sandwich ELISAs) and DNA probe tests (see later). The sequence and nature of the reagents varies from format to format, but the principle is the same.

KEY

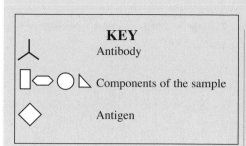

Antibody

Components of the sample

Antigen

Step 1

The assay is carried out in a Test Well. This is coated with an antibody to the antigen of interest.

Step 2

The sample components can interact with the antibody in the Test Well.

Step 3

The antibody recognises only antigen, which it captures. It ignores the other components of the sample.

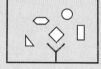

Step 4

The antigen is captured in the Test Well. The other components are washed away.

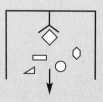

Step 5

Detection antibody is added. This antibody also recognises the antigen and it carries an enzyme.

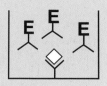

Step 6

This antibody binds to the antigen (if present), making a sandwich.

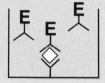

Step 7

The antigen has trapped some of the second antibody in the sandwich. Any unbound antibody (and enzyme) is then washed away.

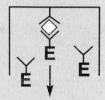

Perfect sandwich

The first antibody has trapped the antigen which has trapped the antibody that carries the enzyme. The Well is now coated with enzyme.

Step 8

The enzyme converts the colourless substrate to the coloured product. The solution in the Test Well becomes coloured.

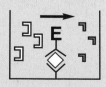

Figure 4 - The end result of a typical sandwich ELISA

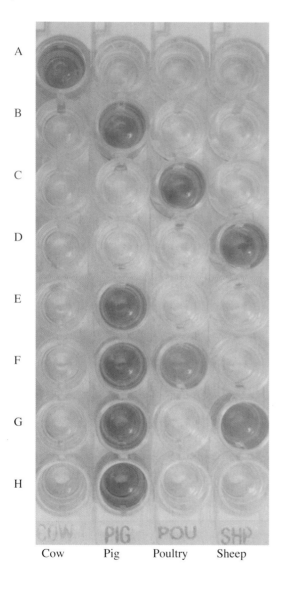

A
B
C
D
E
F
G
H

Cow   Pig   Poultry   Sheep

In this meat species test, samples and standards were analysed for the presence of meat originating from cow, pig, poultry and sheep. The colour in the test wells reveals the presence of meat from the relevant species. The upper four rows (A-D) are standards (controls), providing positive reactions for cow, pig, poultry and sheep respectively (and negative reactions for each of the other three species). The lower four rows (E-H) are test extracts which are shown to contain the following species:

Row E -   pig but not cow, poultry or sheep

Row F -   pig and poultry but not cow or sheep

Row G -   pig and sheep but not cow or poultry

Row H -   pig but not cow, poultry or sheep

Figure 5 - The indirect sandwich ELISA

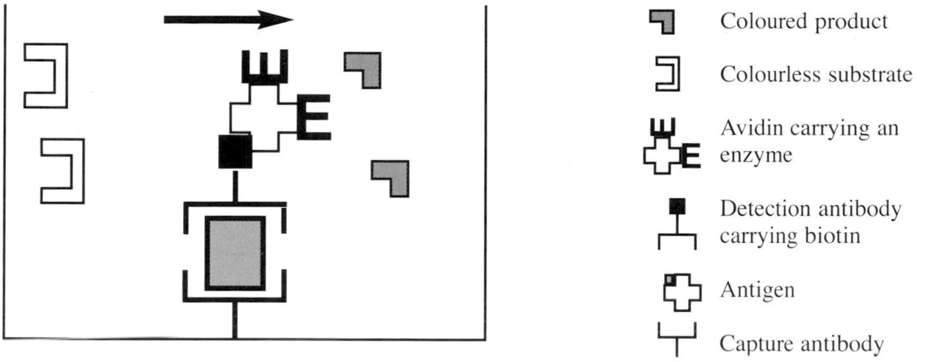

This is similar to the direct sandwich assay (described in Figure 3 and Box 1) except that the capture antibody does not itself carry the enzyme. Instead it might carry some other label (e.g. the small molecule biotin which sticks to the protein avidin). Avidin linked to an enzyme can then be used to detect the 'detection antibody' but only if it has been trapped by the antigen. In effect, there is another layer in the sandwich in this system.

'Dipstick' formats

'Dipstick' type assays often employ the sandwich principle - be it direct or indirect. Again there are variations. In some, for example, in the presence of the analyte, the enzyme becomes trapped on the surface of the dipstick (or paddle) in exactly the same way (i.e. via a sandwich) as it does on the well surface in the sandwich assay. When immersed in a tube of substrate, the enzyme-coated dipstick then generates a change in colour in the solution in the tube.

Latex beads are also used as the signal in some test kits. The antigen and antibody are added, in solution, to a filter paper wick. They are swept along the paper by the flow of the extract, driven by capillary action, and pass over a capture antibody. As they do so, the capture antibody captures the antigen which captures the detection antibody. The latex beads attached to the latter therefore concentrate at the site where the capture antibody was originally attached to the membrane, usually as a tight narrow band. When concentrated in this way, the coloured latex beads form a deeply coloured line.

The robustness of this approach was initially demonstrated in pregnancy test kits designed for home use, in which instance the tests take just a few minutes. Examples of applications to food analysis are listed in Tables 4 and 5, where the tests can sometimes be applied equally rapidly, though in microbiological analysis they do require pre-culturing of the test micro-organism.

Table 4 - Examples of meat species for which there are commercially available dipstick tests

Cow	Pig
Sheep	Goat
Horse	Rabbit
Chicken	Turkey
Kangaroo	Buffalo

Table 5 - Examples of foodborne pathogens for which there are commercially available dipstick based immunoassays

- *E. coli* O157
- *Listeria*
- *Salmonella*

The competitive ELISA

An alternative assay format is the 'competitive immunoassay', which can itself take various forms. In one version (Figure 6), the wells of the multi-well test plate are coated with antigen. Antibody and sample are added to the well simultaneously and incubated. During incubation, the antibody can bind to the antigen on the test-well surface or to that in the sample. The more antigen in the sample, the more the antibody binds to this antigen, and the less antibody binds to the antigen on the plate. Conversely, if the sample contains relatively little antigen then the antibody will bind primarily to the antigen on the test well surface.

In this assay, the colour that develops is inversely proportional to the amount of antigen in the sample. At the extreme, for example, if there is no antigen in the sample then the antibody will bind to the antigen on the plate. The enzyme becomes trapped on the plate and a strong colour develops when the substrate is added (Figure 6).

In contrast, when the sample contains a high level of antigen, most of the antibody will bind to this free antigen, so that much less (if any) binds to the antigen on the test well surface. Consequently much less colour develops (Figure 6). A standard curve can be constructed to relate colour intensity to amount of analyte in quantitative assays (see Figure 7).

Table 6 lists examples of the kinds of food components and contaminants for which there are commercially available competitive ELISAs. This format is often used for analytes which are small molecules, simply because the molecule is literally too small to bind both a capture and detection antibody simultaneously and so cannot be detected in a sandwich assay. However, as can be seen from the table, the format can also used for larger molecules (as exemplified by soya protein).

Table 6 - Food components and contaminants for which there are commercially available competitive ELISA test kits

◆ Soya protein and peanut protein
◆ Bovine caseins
◆ Mycotoxins (fungal toxins) such as aflatoxins, ochratoxin A and zearalenone
◆ Antibiotic residues including amoxicillin, ampicillin, cloxacillin, penicillin G, chloramphenicol and sulphamethazine
◆ Steroids and growth promoters such as clenbuterol and beta-agonists, testosterone, corticosteroids and zeranol.

Figure 6 - The competitive ELISA

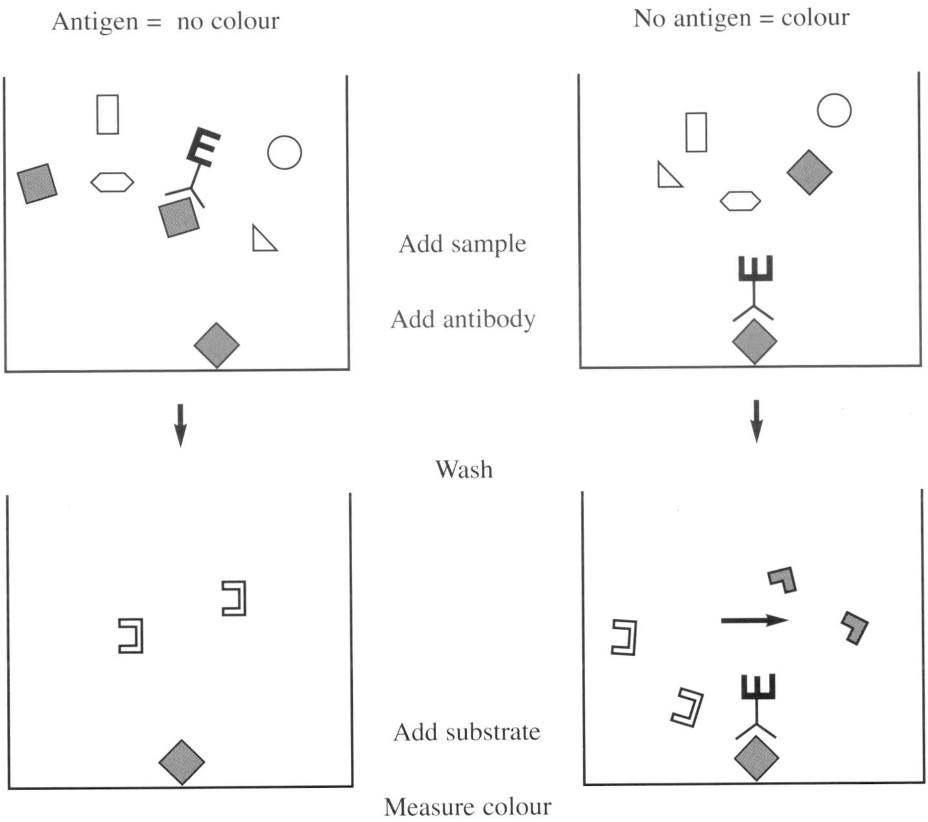

Antigen = no colour

No antigen = colour

Add sample

Add antibody

Wash

Add substrate

Measure colour

The test well is coated with antigen. Antibody and sample are added to the well simultaneously. If no antigen is present in the sample (right) then the antibody will bind to the antigen on the plate. The enzyme on the antibody then catalyses a colour change in the substrate. Absence of antigen in the sample results in the formation of strong colour. Conversely, if the sample contains antigen (left), then antibody will bind to it and less antibody will bind to the antigen on the plate. If the concentration of antibody in the sample is high then no antibody will bind to the antigen in the test well. Subsequently, no colour will develop when the substrate is added.

Figure 7 - Standard curve from a competitive ELISA

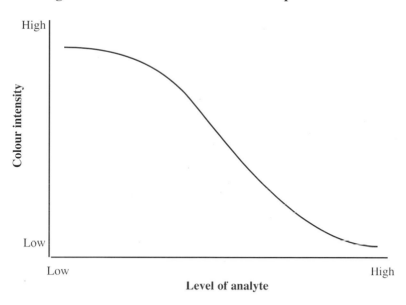

A standard curve is obtained by plotting the intensity of the colour (absorbance) against the amount of analyte. By reading the absorbance of a sample and using the linear part of the graph, the level of the analyte can be determined.

Common variations on the competitive ELISA

Again there are many variations on the competitive ELISA format. For example, in some assays, the antibody may be attached to the test well surface, so that it is the antigen in the sample extract which 'partitions' between the antibody on the well surface and the antibody free in solution. There are also options for both direct and indirect formats (as with the sandwich ELISA).

Immuno-diffusion

This is an example of an assay where the antibody-antigen complex is visualised directly to obtain the test result - there are no enzymes or colour change reagents involved. It is also an example of an assay which relies on the ability of the antibody to bind to two different molecules of the same antigen at the same time.

The antibody is added to a localised well or disk on an agar gel. Sample extract is added to a similar spot some distance from the antibody. Both the antibody and the sample extract move through the gel by simple diffusion. If antigen is present in the extract, where the antibody and antigen meet at appropriate concentrations, they bind together to form a large molecular complex. This complex - or net of molecules - is so large that it is no longer soluble. It precipitates to form a white band in the agar gel (Figure 8).

This precipitation only occurs if the antibody and its complementary antigen are both present. Just like the ELISA then, this immunoassay can be used as a specific test for the presence of antigen.

Figure 8 - The principle underlying immunodiffusion tests

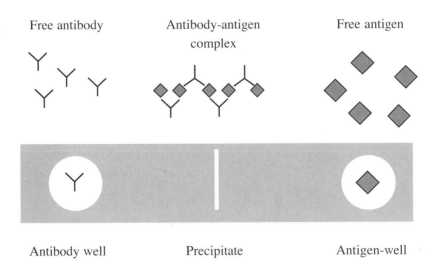

Antibody is added to one well in an agar gel, and sample to a second well. Both diffuse through the gel. If the sample contains antigen recognised by the antibody, then where the two meet they will form a huge molecular complex which precipitates. This is visible as a white band.

There are several commercially available immuno-diffusion test kits for use in food analysis (Table 7). Immuno-diffusion tests are simple to use and quick to set-up - literally just a minute or so for each sample if the extracts, reagents and gels are prepared in advance. However, they are also relatively slow - typically taking 24 hours to generate results - as they rely on passive diffusion of the components through the gel.

Table 7 - Food analyses for which there are commercially available immunodiffusion test kits

- ◆ Confirmation of species of origin of beef
- ◆ Confirmation of species of origin of poultry

Latex agglutination

Like immuno-diffusion assays, agglutination reactions rely on the ability of an antibody to bind two molecules of the same antigen at the same time. In latex agglutination tests, the antibodies are also coupled to coloured latex beads. As the antigen forms bridges between the antibodies on different beads (Figure 9), a complex three-dimensional mesh of bead, antigen and antibody forms. This agglutination reaction is fast and clearly visible, and forms the basis of test kits for several food-borne bacteria (Table 8).

Table 8 - Examples of micro-organisms for which there are commercially available kits based on latex agglutination

- ◆ *Bacillus cereus* diarrhoeal toxin
- ◆ *Cumpylobacter*
- ◆ *E.coli* O157
- ◆ *Listeria*
- ◆ *Salmonella*
- ◆ *Staphylococcus aureus*
- ◆ *Staphylococcus aureus* enterotoxin

Figure 9 - The principle of latex agglutination

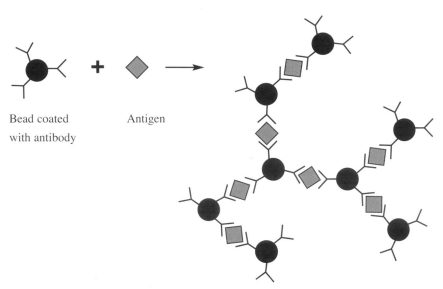

Bead coated
with antibody

Antigen

The latex beads are coated with antibody. If the corresponding antigen is present in the test sample, then it will bind to the antibody. In direct analogy to the sandwich ELISA (see Figure 3 and Box 1) the antigen will get sandwiched between the antibody on the beads, so that the beads will cluster (agglutinate). This agglutination is visible to the naked eye.

Immuno-affinity columns

These columns are sometimes used as tests in their own right, but more often are used to prepare samples for further analysis by other methods (either immunoassay or 'conventional' chemical analysis). Explanation of the mechanics of this format reveals why this is so.

The antibody is coated on to the surface of beads which are packed into a small column (Figure 10). The sample extract is passed through the column. As it passes through, the antibody on the beads literally plucks the antigen from the sample - assuming antigen is present. The remainder of the sample passes through the column, leaving the antigen bound to the column. The antigen can be released from the antibody's grip by washing the column through with a suitable buffer; this resulting solution contains the antigen without many of the impurities of the original extract.

The antigen is then detected more easily by conventional methods (e.g. liquid chromatography) or in an ELISA (Table 9). One interesting variation on this in the analysis of aflatoxins - toxins produced under certain circumstances by the fungus *Aspergillus flavus* in some cereal products. Aflatoxins naturally fluoresce under ultra-violet light. Aflatoxin trapped on an immuno-affinity column can be eluted (washed-off) with an appropriate buffer and trapped in a small tip which, when placed under UV light, will fluoresce if aflatoxin is present.

Table 9 - Examples of food contaminants for which there are immuno-affinity column systems for food analysis

Mycotoxins such as aflatoxins and ochratoxin A

Growth promoters such as clenbuterol and zeranol

Figure 10 - The principles of immuno-affinity columns

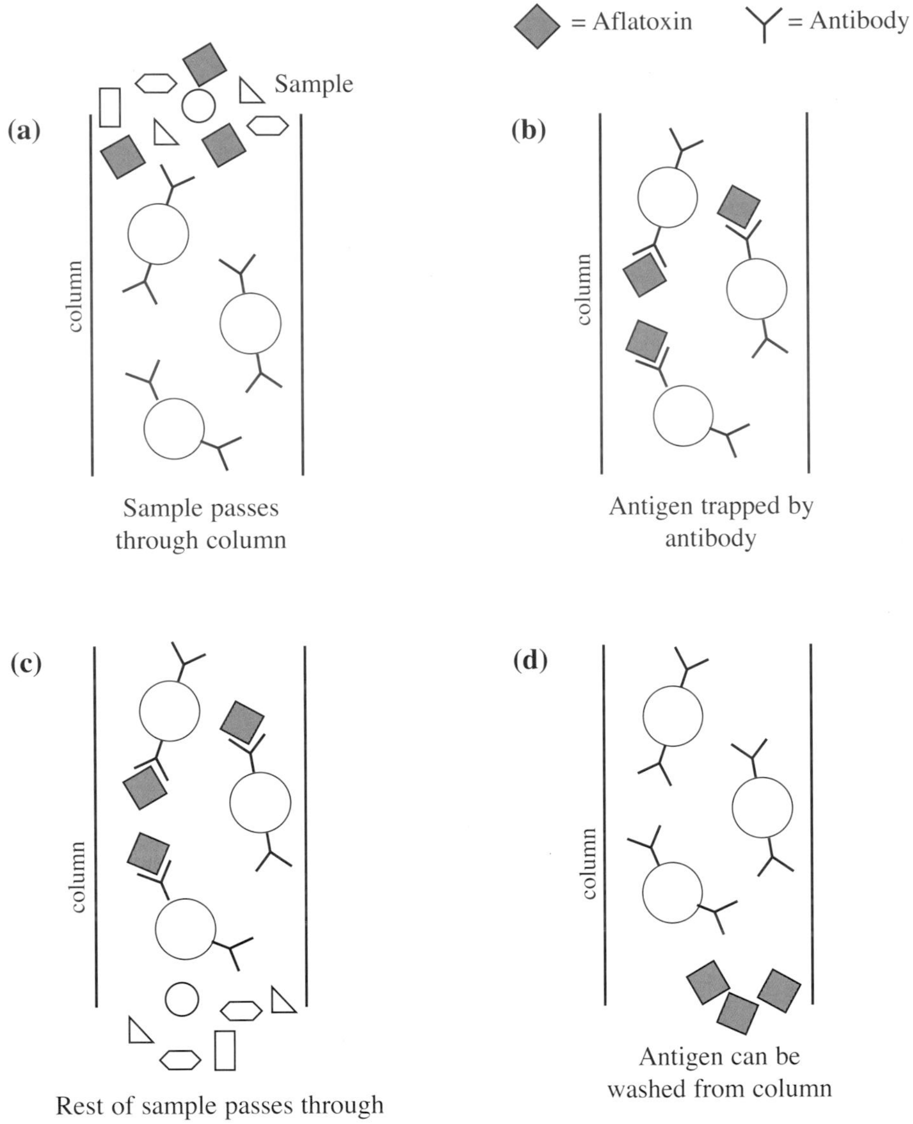

◆ = Aflatoxin   Y = Antibody

(a) Sample

column

Sample passes
through column

(b)

column

Antigen trapped by
antibody

(c)

column

Rest of sample passes through

(d)

column

Antigen can be
washed from column

The sample is passed through the column (a). Antibody immobilised on the beads in the column traps antigen (b) as the sample passes through the column. When all the other sample components have washed through the column (c), the antigen can be washed off the antibody (with an appropriate buffer) (d), and analysed separately.

Immuno-magnetic beads

Immuno-magnetic beads provide a variation on this theme and are used in microbiological analysis. The magnetic beads, coated with antibody to the bacterium of interest, are mixed with the test culture. If present, the bacterial cells will bind to the antibody on the bead surface. The beads can be collected by application of a magnetic field, and the unbound cells washed away. The bound microbes are then detected by a conventional microbiological plate count (Table 10) in a process which can be fully automated.

Table 10 - Magnetic bead systems

Immunocapture is used to isolate the bacterial cells from a culture, and the cells are then detected by conventional plate counting techniques.

- *E. coli* O157
- *Listeria*
- *Salmonella*

2.3 Immunoassays - advantages and limitations

Immunoassays offer many advantages to the analyst. They can be relatively quick, simple, sensitive (i.e. detect low levels of chemical analytes) and reliable. Also, because the same test format (e.g. a sandwich ELISA in a 96-well test plate) can be applied to a wide range of analytes, immunoassays can be highly automated, allowing high volume processing of many samples in parallel.

However, there are clear limitations to their applicability. First, immunoassays are only of use when looking for a specific analyte. For example, using an assay to test for pork will tell you whether pork is present, but separate tests will need to be used for other meat species, soya, cereal proteins and so on. The same applies for detecting different foodborne pathogens. Second, they cannot be used where there are not clearly defined and consistent differences between samples. For this reason they cannot, for example, be used to distinguish samples which are biologically the same but which differ in geographical origin (e.g. Welsh lamb from New Zealand lamb). Third, and on a practical level, most immunoassays must be carried out with great care to avoid problems of cross-contamination which can give misleading results. Fourth, although the assays themselves can be rapid and set to standard formats, sample preparation varies considerably from analyte to analyte and, for a single analyte, from food matrix to food matrix. Sample preparation is therefore often the limiting step.

Finally, although several thousand immunoassays have been developed for food analysis, because of the costs of converting research-based tests into robust, reliable and easy-to-use test kits, only a proportion of these are available as commercial kits for use in routine testing laboratories.

2.4 Re-cap of the basic principle

Immunoassays are all based on the specific interaction of an antibody with its antigen. They also rely on an effective detection system and usually require a solid phase. These ingredients can be brought together in a bewildering array of assay formats. Examples described include the sandwich ELISA, competitive ELISA, latex agglutination, immuno-magnetic beads, immuno-diffusion, and immuno-affinity columns. Within most of these there are many subtle variations. The key point, however, is that immunoassays can be used to detect individual components of complex mixtures - including food components, contaminants (chemical and microbiological) and adulterants - quickly, simply and reliably, by exploiting the highly specific interaction between antibody and antigen.

3. DNA METHODS

DNA-based methods have revolutionised aspects of diagnostic technology in the clinical and forensic fields. Their incredible power in detecting and characterising organisms, and material derived from organisms, coupled to enormous improvements in their user-friendliness, has led to fascinating applications of direct relevance to the food industry. Examples so far include the detection and typing of food-borne pathogens, authenticity testing of meat and fish products, and detection of genetically modified food materials. This section outlines the key principles on which all DNA methods are based, starting with a brief description of the structure of DNA. Specific examples of where the methods are in common use are tabulated, and some of these elaborated upon in the case studies. For general interest, and to illustrate the emergence of the technology, some landmark examples of DNA analysis in other fields are briefly highlighted.

3.1 The structure of DNA

DNA (deoxyribonucleic acid) is an extremely long molecule, but relatively simple to understand (Figure 11). Often referred to as the code of life, its structure reflects its simple four-lettered alphabet: through variations in the sequence of these four letters DNA carries all the information needed to construct each living organism.

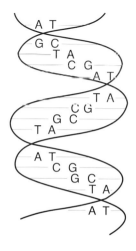

Figure 11 - The structure of DNA

DNA is a double helix of two complementary polynucleotide chains, held together by weak bonds between adenine (A) and thymine (T), and between cytosine (C) and guanine (G). The two chains intertwine in a double helix. The sequence of the bases A, C, G and T carries the genetic information of the organism. The figure shows a short stretch. DNA typically runs for millions of base pairs.

The four letters of the DNA alphabet are A (for adenine), C (cytosine), G (guanine) and T (thymine). Adenine, cytosine, guanine and thymine are strung together on a sugar-phosphate backbone to make up a polynucleotide chain. The base, attached to one sugar molecule (deoxyribose) and phosphate group, makes up a mono nucleotide. When strung together, the mono nucleotides make a polynucleotide chain. DNA is composed of two polynucleotide chains intertwined in a double helix (Figure 11). The two chains are not identical but complementary: the A of one chain lies opposite the T of the other, and C lies opposite G. Weak bonds (hydrogen bonds) between A and T, and between C and G, hold the two chains together in the double helix.

Genes are stretches of DNA which carry specific information written in the four lettered genetic alphabet. Some genes occur only once in the entire genome (the genome is the full set of genes in an organism) whereas others may occur in several or many copies. In addition, an organism's DNA will contain many sequences which appear to carry no information (sometimes repeated as thousands of copies) and which is often referred to as 'junk' DNA. For DNA based tests it often does not matter whether the DNA carries information or is junk DNA - simply whether or not differences can be observed between the DNA of the organisms being compared.

3.2 DNA methods - the main components

DNA methods all employ one or more of three basic components - DNA probes, DNA cutting, and DNA amplification. As with the immunoassays described in Section 2, the assays also usually require a solid phase and a signalling system. Each of these is now briefly described as a prelude to describing how the ingredients can be combined in various ways to create different assay formats.

DNA probes and how they work

A DNA probe, or gene probe, is itself a piece of DNA, but it is a single strand rather than a double helix (Figure 12). Each probe can be used to reveal the presence of (i.e. to probe) a specific 'target' DNA sequence which may be tucked away amongst the rest of the organism's DNA (genes). Rather than recognise the double helix, the

Figure 12 - Stylised structure of a DNA probe

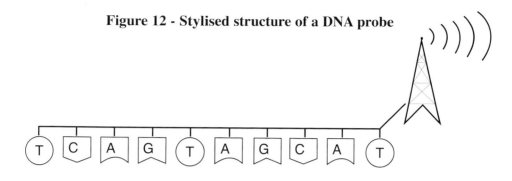

DNA probes are single strands of DNA, usually from around 100 to a 1000 bases long. They carry a signal (or label - depicted by the radio tower) so that the user can monitor whether the probe has bound to its target. Traditionally the signal has been radioactive phosphorus (incorporated into the DNA backbone) but now a range of enzyme-based systems are available, allowing detection on the basis of colour changes or light emissions (as with the ELISA formats).

Figure 13 - Probe-target hybridization

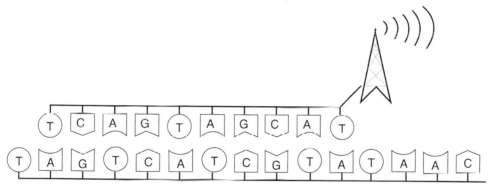

On the principle of complementary base pairing (A-T and C-G), probes hybridize with complementary target sequences to form a stable double helix (helix omitted for clarity). The label (e.g. enzyme or radioactive marker), depicted by the radio-tower, 'reports' the presence of the probe (and hence the target sequence) to the investigator. Note that in practice probes are at least 20 nucleotides long, and may be several thousands of nucleotides long.

Figure 14 Cutting DNA with a restriction enzyme

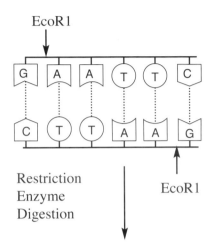

Restriction enzymes break the 'backbone' of the polynucleotide chains, thus fragmenting DNA. Each restriction enzyme cuts at a specific site in the DNA molecule, and this is determined by the sequence at that site. For example, EcoR1 (which is shown) will cut the DNA at all the sites where the sequence GAATTC occurs. There are now several hundred known restriction enzymes.

probe recognises one of the two polynucleotide strands - it sticks to this strand by virtue of the 'complementary base pairing' - between A and T and between C and G, as described above.

For example, a probe with the sequence TCAGTAGCAT would hybridize with (i.e. recognise and stick to) the target sequence AGTCATCGTA (Figure 13). A probe carries a 'label' which tells the investigator that it has found and hybridized to its target sequence (Figure 13). The label can be anything that is easy to detect - an enzyme that causes a colour change, a fluorescent molecule or a radioactive isotope.

Cutting DNA - the use of 'restriction enzymes'

It is possible to cut the DNA double helix at specific sites, using biochemical scissors called restriction enzymes. A typical example is the enzyme *Eco*RI, as shown in Figure 14. This cuts DNA where it sees the sequence GAATTC but not at any other sequence. There are many other enzymes which can be used to cut DNA, and each will recognise a particular DNA sequence and so cut the double helix at different points.

DNA amplification - the polymerase chain reaction

It is possible to make many copies of a piece of DNA biochemically, using a DNA amplification procedure such as PCR (the polymerase chain reaction). There are now many variants of PCR but they are all based on the same basic concept and yield the same result: using PCR it is possible to make millions of copies of a piece of DNA in a tube in just a few hours. PCR works by mimicking nature's way of making DNA, in which the two polynucleotide strands of the double helix separate, to act as templates for the synthesis of two new complementary strands.

The ingredients in a PCR tube are the target DNA (to be amplified), the polymerase enzyme, the nucleotide building blocks (G, A, T and C, that make up the polynucleotide strands) and the 'primers' (Figure 15). Primers are short pieces of

Figure 15 - The polymerase chain reaction (PCR)

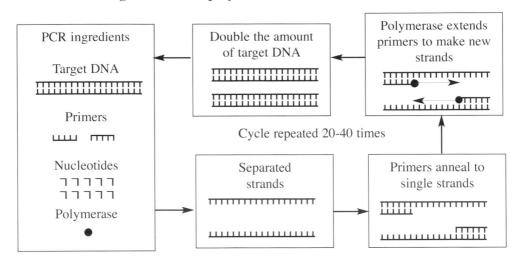

Multiple copies of DNA can be made using PCR. The target DNA is mixed in a tube with primers, nucleotides and polymerase. The tube is heated, to separate the two strands, and then cooled so that the primers can pair with the regions of the strands that they complement. The polymerase then runs along the target strand (which acts as a template) stringing new nucleotides on to the primers to make new strands. Theoretically each run of the cycle doubles the number of strands and even though the reaction is not as efficient as this in practice, a run of 30-40 cycles will make more then one million copies of the original.

single stranded DNA complementary to the sequences which flank the region to be amplified. They tell the polymerase where to begin making a new strand, and so they define the sequence to be amplified. The tube of ingredients is heated in order to separate the target strands. Slight cooling then allows the primers to hybridize with their complementary regions of the two strands. The polymerase then begins to make two new strands - it uses the target as a template and strings nucleotides on to the primers until a new strand has been made. Then the heat is re-applied to drive apart the new strands and start the cycle again. Twenty cycles results in a million-fold amplification of the specific sequence of interest - in other words, over a million copies of the same DNA fragment are generated.

RNA - an added dimension

RNA is another naturally occurring nucleic acid found in all living cells. It is increasingly being exploited in some molecular methods used in food analysis. At least some of the tests routinely applied to DNA can also be applied to RNA, sometimes with added advantages.

Box 2 - Where does RNA fit in?

There are three kinds of RNA in a cell: messenger RNA (mRNA), transfer RNA (tRNA) and ribosomal RNA (rRNA). They each play a role in the production of proteins in cells. Proteins are made of 'building blocks' called amino acids - all proteins are long chains of amino acids. Different amino acid sequences lead to proteins with very different properties - consider, for example, the different properties between soya protein (a seed storage protein) and muscle protein.

It is the information in the DNA code that determines which amino acid comes where in the chain, and therefore determines the properties of proteins. There are two main stages in decoding DNA and making proteins. First, a copy of the DNA code is 'transcribed' (in a process called transcription) to create each of the three types of RNA. The 'code' in the messenger RNA (made up of A,C,G,U) is then 'translated' into the corresponding amino acid sequence (in a process called translation). This is illustrated in the figure. The ribosomal RNA is a major structural component of the ribosomes - the parts of the cell where proteins are made. The transfer RNA is involved in transferring the amino acids to the messenger RNA on the ribosomes, so that each amino acid is slotted into the sequence in the right order.

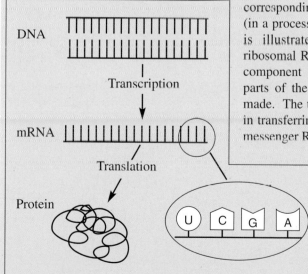

RNA stands for ribonucleic acid. There are in fact three types of RNA - called mRNA, tRNA and rRNA - each of which performs a different role within the cell (see Box 2). It differs from DNA in several ways:

- Although it is a polynucleotide chain, it is usually a single strand and not a double helix of two intertwining strands. Also, the sugar in the sugar-phosphate backbone of the chain is ribose in RNA (as opposed to deoxyribose in DNA).
- Three of the bases ('letters') that make up RNA are the same as DNA (A,C and G) but in RNA the fourth base is uracil ('U') rather than thymine ('T') so that an RNA molecule has sequences based on ACGU
- Compared to DNA, which is millions of nucleotides long, RNA molecules are shorter - ranging from tens to thousands of nucleotides long

Despite the differences between DNA and RNA, especially the different bases used, a single strand of DNA can and will hybridise with a complementary strand of RNA. In this interaction, C still pairs with G, but the U in the RNA pairs with the A in the DNA. This means that a DNA probe can be used to detect an RNA target (see Figure 16). One of the main advantages of analysing RNA is that the cell will usually hold many copies, whereas a DNA target may be present at only a single copy per cell, potentially giving the assay greater sensitivity.

Figure 16 - A DNA probe hybridised to an RNA target

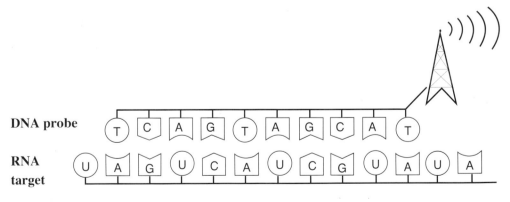

RNA is a single stranded nucleic acid in which the bases are adenine (A), uracil (U), cytosine (C) and guanine (G). A DNA probe can attach to an RNA target in the same way as it attaches to a DNA target - by complementary base pairing (see Figure 13). In this case however, the A of the DNA probe complements the U of the RNA target. Again, in practice, the probes and targets are usually much longer than depicted.

DNA evidence comes to court

Following the discovery of genetic fingerprinting (RFLP - see p.37) in 1984, it quickly became apparent that there were many potential practical applications including forensic analysis (1). The first reported use of DNA fingerprinting (RFLP) in a murder case was in 1987 (Narborough, UK) (2). An initial suspect was released when it was established that his DNA did not match that of the forensic samples obtained from the victim. With evidence that the perpetrator was local, the police, with the support of the community, conducted a mass-screening of all 5,500 of the district's suspect population - initially by blood group and subsequently (on those who could not be excluded in this way) by genetic fingerprinting (2). Despite initially evading the mass-screen, the killer was identified following an overheard pub conversation about the screening exercise, and his DNA was found to match that found at the scene. He confessed both to this murder and an earlier one and was sentenced to life imprisonment.

With the discovery and application of PCR, it also quickly became established as a tool for the forensic scientist. The first case involving PCR-derived forensic evidence in a British court involved the identification of a murder victim from skeletal remains several years after she was first reported missing (3). Having been discovered by workmen in a garden, the approximate age of the victim was established from dental records. Meanwhile a medical artist used facial reconstruction to create a plaster-cast likeness, a picture of which was released to the public. Within two days a tentative identification had been made, which was strongly corroborated by dental records. However, using PCR-based analysis of DNA extracted from the bone it proved possible to confirm the identity of the remains by comparison with DNA samples from living relatives (3).

References:

1. Connor, S. (1988) Genetic fingers in the forensic pie. New Scientist 28 January 1988 p31-32.

2. Kirby, L. (1990) DNA Fingerprinting: An Introduction. Stockton Press. ISBN 0 333 54024 7

3. Davies, K. (1991) Untitled note. Nature 352 (1 August 1991) p381.

Molecular archaeology

The field of 'molecular archaeology' involves the application of DNA analysis to fossils and other archaeological materials. Some of the oldest DNA extracted is from fossils of plant origin or animal material encapsulated in amber (1). DNA has been successfully extracted and typed from fossilised leaf of a Miocene *Magnolia* specimen (around 20 million years old) (2) and from amber-preserved leaf of the extinct tree *Hymenaea protera* (3), for example. Such DNA analyses are helping to piece together the evolutionary relationships between extinct and living plant species. In the context of human archaeological remains, they are also being used to confirm and extend the findings derived from conventional studies of archaeological sites - through establishing the sex of individual specimens, the familial relationships between individuals at particular sites and even tracking the migration of specific populations (1).

References:

1. Brown, T.A. and Brown, K.A. (1994) Ancient DNA: using molecular biology to explore the past. BioEssays 16 719-728.

2. Golenberg, E.M., Giannasi, D.E., Clegg, M.T., Smiley, C.J., Durbin, M., Henderson, D. and Zurawski, G. (1990) Chloroplast DNA sequence from a Miocene *Magnolia* species. Nature 344 656.

3. Poinar, H.N., Cano, R.J. and Poinar, G.O. (1993) DNA from an extinct plant. Nature 363 677.

3.3 DNA detection and typing techniques - the main variants

The above components - namely DNA probes, DNA cutting and DNA amplification - can be used alone or in various combinations to provide an extremely powerful set of diagnostic tools. The different assay formats yield different information with differing degrees of rapidity. Unlike many of the immunoassay systems described above, these assays are generally qualitative (i.e. detect presence / absence) rather than quantitative.

The main techniques are described here, with some description of their general features in practical terms (e.g. speed, simplicity, reliability), accompanied by some examples of commercially available systems for food analysis. However, as with immunoassays, new variations on the basic themes, combining the ingredients in new and innovative ways, are frequently published.

Probes - dot-blots assays

The simplest format for a DNA probe test is the 'dot-blot' assay, in which target DNA is immobilised on a membrane as a dot. The two strands of the double helix are first separated (by heating, for example) to generate single stranded DNA, in which the exposed bases (A, C, G and T) are free to hybridize with the probe.

This single stranded DNA is immobilised (e.g. by baking) on a membrane of nitrocellulose or nylon, preventing the complementary strands of the original helix from re-annealing. The probe applied to the membrane under appropriate conditions hybridizes with its target DNA, excess (unbound) probe is washed away, and the presence of the target is revealed by the probe label. If the probe recognises the target (i.e. if the target sequence of interest is present), the label reveals it as a dot (Figure 17 - Positive). If the target sequence is not present, then no dot develops (Figure 17 Negative).

Figure 17 - Results of a dot-blot assay

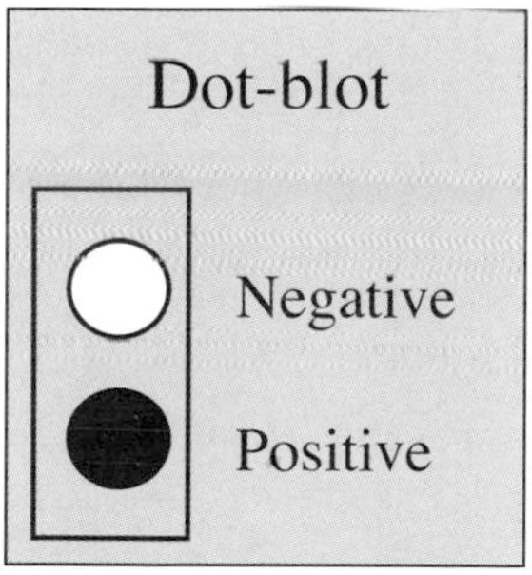

The target DNA is immobilised on a nylon membrane as a dot and then challenged with a probe. If the probe hybridises with the target sequence (i.e. recognises and sticks to it) then the signal is revealed as a circular blot on the membrane (or photographic image of the membrane).

If the signal used is an enzyme-mediated colour change, similar to that used in membrane based immunoassays, then the dot appears as a coloured stain on the membrane itself. If, however, the signal is light emission or radioactive phosphorus, then the signal is detected by exposing the membrane to photographic film, which 'fogs' when exposed to light or radioactivity.

Most probe-based assays take 1-2 days to complete, involve the use of laboratory equipment and require skilled personnel.

Beads, paddles and dipsticks

Some formats use 'paddles', dipsticks or beads as a solid phase. Figure 18 depicts an assay using a bead system, but the principle is exactly the same if the probe is attached to a paddle, dipstick or the surface of a tube or test-well. The bead carries a 'capture' probe. Also present is a standard labelled probe. These two probes are complementary to two contiguous regions of the target DNA (or, in some cases, target RNA). Two hybridisation reactions take place - one attaching target to the bead, the other attaching probe to the target. The beads can be collected (e.g. by centrifugation or magnetic separation) to allow washing steps for separation of bound and free label and removal of excess test sample.

In this type of assay, the signal is usually an enzyme-mediated colour reaction, with the development of a soluble coloured product in the solution surrounding the paddle or bead. Although the probe can be labelled directly with enzyme, it is more usual for it to be labelled with a small molecule (e.g. biotin) which can be detected using avidin or an antibody, which itself is coupled to an enzyme.

Use of such solid phases has made the development of relatively rapid and simple test kits feasible, significantly reducing the need for the manual dexterity required to perform dot-blot assays which traditionally are lengthy and fiddly (Table 11).

Figure 18 - The principle of a 'bead capture assay'

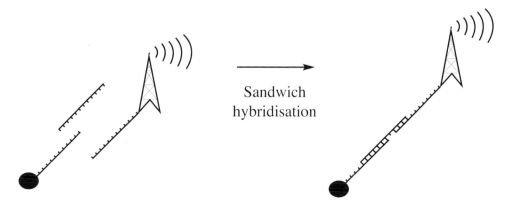

The bead carries a 'capture' probe. A standard labelled probe is also present. The two probes are complementary to two contiguous regions of the target DNA (or in some cases, a target RNA). Two hybridisation reactions take place: one attaches the target to the bead, the other attaches the probe to the target. The beads can be collected (e.g. by magnetic separation or centrifugation) to enable the separation of bound and free label. In some cases beads are replaced by paddles or tubes.

Table 11 - Examples of foodborne micro-organisms for which there are commercially available DNA probe assays

Analyte/test	Format
Campylobacter	Hybridisation protection assay DNA/RNA probe sandwich assay (paddle)
E. coli (including O157)	DNA/RNA probe sandwich assay (paddle)
Listeria monocytogenes	Hybridisation protection assay DNA/RNA probe sandwich assay (paddle)
Salmonella	DNA/RNA probe sandwich assay (paddle)
Staphylococcus aureus	Hybridisation protection assay

Hybridisation protection assay

A further refinement eliminates the solid phase completely. In the format known as the hybridisation protection assay (HPA), the probe is labelled with an acridinium ester which can react with hydrogen peroxide to release light (in a chemiluminescent reaction). In the HPA format, the probe is allowed to hybridise to its target DNA or RNA sequence. Once hybridised, the probe and its label (the acridinium ester) is protected from a hydrolysing agent which selectively destroys the acridinium ester attached to any unhybridised probe (Figure 19). Any remaining chemiluminescence, as measured with a luminometer, is therefore generated by the hybridised probe, and is indicative of the presence of target nucleic acid. Table 11 lists examples of foodborne pathogens for which there are commercially available tests based on the HPA.

Figure 19 - Principle of the hybridisation protection assay

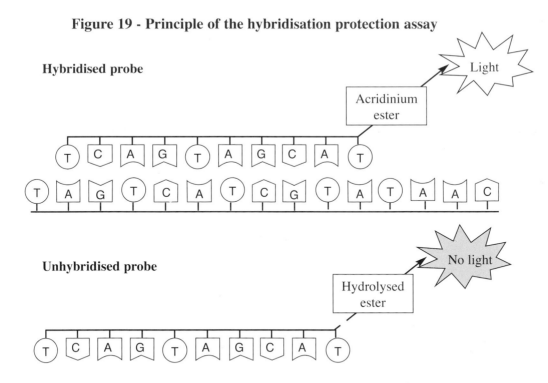

The probe is labelled with an acridinium ester which can generate light in a chemiluminescence reaction. The ester attached to unhybridised probes is destroyed by a hydrolysing agent whereas the ester attached to the hybridised probe survives. Emission of light therefore indicates the presence of target sequence.

DNA profiling - RFLP

RFLP stands for restriction fragment length polymorphism, and RFLP analysis forms the basis of conventional 'genetic fingerprinting'. The principle of RFLPs is illustrated in Figure 20. It involves the use of restriction enzymes and DNA probes but not DNA amplification.

Restriction enzymes cleave DNA at the site of a short specific sequence of nucleotides, resulting in fragments of particular sizes. The fragments are then separated on the basis of size using electrophoresis. A sample containing the fragments is deposited in a small well in a gel and an electric current applied. As DNA is negatively charged, it is drawn towards the positive pole of the gel. Smaller fragments, which pass through the 3-D mesh of the gel more easily, move through

Figure 20 - The principle of RFLPs
(restriction fragment length polymorphisms)

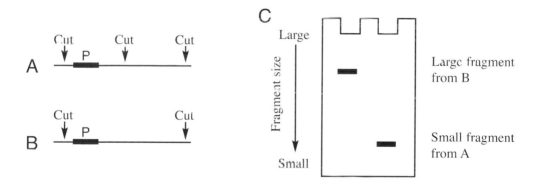

The DNA double helix is cut at specific sites (marked 'cut') by restriction enzymes (RE), each RE recognising a specific DNA nucleotide sequence. A simple change in sequence (as occurs naturally) can result in the loss of an RE site (compare A with B). Digestion with the RE produces DNA fragments of different sizes, the smaller migrating further than the larger on an electrophoretic gel. After transfer of the DNA fragments to a membrane (a procedure called Southern blotting - after its inventor, E.M. Southern) the probe (marked 'P') highlights RFLPs as different banding patterns (C), usually with multiple bands.

the gel more quickly than larger fragments, which have to squeeze their way through. Once separated, the fragments are transferred to a nylon membrane (which is simply pressed against the gel to make a facsimile), attached to the membrane (e.g. chemically or baked on) and then detected with a DNA probe in the same way as in a dot-blot assay.

Polymorphisms (different forms) arise where nucleotide changes have arisen at one of the restriction sites: the enzyme is unable to cut the DNA at that site, so that the resulting fragments are of different sizes (Figure 20). Different DNA sequences (e.g. from different strains of a micro-organism) will cut in different places, and so exhibit different RFLP patterns (called 'banding patterns').

There are no set rules about which probes or restriction enzymes have to be used, and in fact, by varying the combinations of probes and enzymes, the technique offers considerable flexibility and can give extremely good discrimination. By trial and error the best combinations of probes and enzymes for a particular application can be identified. Combinations that give good discrimination for one application may not give the discrimination required in other instances.

On a practical level, RFLP takes several days to carry out and requires the use of sophisticated equipment by skilled personnel in a specialist laboratory. However, as a laboratory-based technique it has enormous power, as witnessed by the now routine and widespread use of genetic fingerprinting in forensic analysis and clinical diagnostics.

DNA profiling - Ribotyping

One example of DNA profiling that is proving particularly useful for identifying microbial food contamination is 'ribotyping'. This is a basically RFLP analysis of a specific set of microbial genes, which carry the information for ribosomal RNA - hence the term ribotyping. The technique has now been automated so that it can be used as a routine tool to generate comparative DNA profiles of specific strains of microbes in as little as 8 hours (see Table 12 and Box 3).

Table 12 - Examples of foodborne micro-organisms to which automated ribotyping has been successfully applied

Pathogens

- *E. coli*
- *E. coli* O157
- *Listeria* spp.
- *Salmonella* spp.
- *Staphylococcus aureus*

Others

- *Pseudomonas* spp.
- *Burkholderia* sp.
- *Stenostrophomonas* sp.

Box 3 - Case study

Tracing bacterial contaminants

An automated ribotyping system, called the RiboPrinter™, allows the rapid characterisation of problem microorganisms. It exploits the similarities between the genes that code for ribosomal RNA in bacteria: the sequences of these genes are highly conserved but the genes themselves vary in number and position within the chromosomes of different bacteria. To ribotype a bacterial isolate, the DNA is extracted from a colony, digested with a restriction enzyme and the resulting fragments separated by electrophoresis. Differences in the gene number and position will create differences in the position of the fragments on the gel. The fragments are transferred to a membrane, immobilised, and detected with a chemiluminescent probe, revealing the fragment patterns which can be photographed (see below). Automation of ribotyping in the RiboPrinter™ allows rapid (8 hours) typing of isolates and facilitates direct comparison of the fragment pattern obtained with a database of those of many thousands of bacterial isolates for rapid comparison and identification.

In one case, a ready-to-eat food was found to be contaminated with *Staphylococcus epidermidis*, using conventional microbiological analyses. This organism was found in several areas of the production site, but it was not possible to conclude which was providing the reservoir of contamination. Potentially, the whole site would need to be sanitised. Using RiboPrinter technology, all of the isolates were typed. Only one - isolated from the hands of a production operative - was found to match that from the contaminated product (see figure). Improved

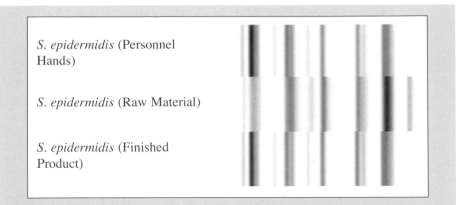

S. *epidermidis* (Personnel Hands)

S. *epidermidis* (Raw Material)

S. *epidermidis* (Finished Product)

personal hygiene measures were introduced and the problem was eliminated, avoiding costly closure of the site and reduced factory output.

A similar problem, but with regard to *Listeria* contamination of a ham sandwich, was solved using riboprinting. In this case, a *Listeria* isolate in the product was matched with that on the shoes of production personnel and the handle of a slicer, which could be distinguished from other potential sources of *Listeria* and targeted for remedial action.

In addition to pathogens, the technique can be applied to spoilage organisms such as *Pseudomonas*. The approach was used to solve a problem for a food company which had been experiencing product spoilage. Conventional microbiological analysis implicated *Pseudomonas*, but could not pinpoint the source. Using RiboPrinting™ an exact match was found between the spoilage strain and that found in condensation on processing equipment. Further investigation revealed that the condensation had arisen because of changes in plant layout which had reduced airflow. Having identified the cause, remedial action could be taken to resolve the spoilage problem.

References:

1. Anon (2000) Timely, definitive bacterial identification and tracing of a contamination in just 8 hours. CCFRA Newsletter, January 2000, p2.

2. Broomfield, P.L.E. (1996) Molecular typing of bacterial isolates as a contract service. Microbiology Europe 4 (4) 24-26.

3. Ridley, A. (1998) Bacterial food spoilage: characterisation of key strains by ribotyping. CCFRA Research Summary Sheet 1998/60

PCR, multiplex PCR (mPCR) and reverse transcriptase PCR (rtPCR)

PCR can form part of a single detection system (e.g. amplifying a DNA sequence for detection with a probe or by electrophoresis) and as an integral part of profiling systems (see later). In standard PCR, primers are used to amplify the target sequence(s) of interest. The target can then be separated from other DNA on an electrophoretic gel and detected by staining. This system has been applied to the detection of *Salmonella*, for example (see Box 4).

Multiplex PCR takes this a step further. It involves using mixtures of primers for quite different DNA targets, to carry out simultaneous amplification of multiple targets in the same tube at the same time. Although this brings potential savings in terms of reagents, time and costs, and can also reduce the tube-to-tube variation in comparative analyses, it can be difficult to establish PCR conditions under which all the sets of primers perform well.

A good published example applying this approach to a model system is the use of multiplex PCR (mPCR) for simultaneous detection of *Escherichia coli* O157, *Salmonella* Typhimurium and *Listeria monocytogenes* in a culture. This is termed 'Non-linked mPCR' because the target sequences derive from three different genomes. In contrast, 'Linked mPCR' involves amplification of three target sequences from the same genome.

A completely different variation is rtPCR (reverse transcriptase PCR). In the same way that RNA is a copy of DNA (see Box 2, page 29), the enzyme reverse transcriptase can be used to make a DNA copy of an RNA sequence. This DNA copy can then be amplified by standard PCR. This system of rtPCR is used widely in research but not in routine food analysis.

Table 13 - Examples of foodborne micro-organisms for which there are commercially available PCR based methods

- *Listeria*
- *Salmonella*
- *E. coli* O157

Box 4 - Case study

Detecting *Salmonella* in food by PCR

When testing for bacterial pathogens, time is of the essence in terms of both healthcare and economics. With rising cases of salmonellosis, food and clinical microbiologists have been quick to explore the speed and sensitivity of PCR. One of the first commercially available PCR-based tests for *Salmonella* in food - called BAX™ - is typical in that it can help circumvent the need to extensively culture the bacterium prior to analysis. It provides a definitive result within just 28 hours - compared to several days by conventional analysis.

The potentially low infective dose of *Salmonella* means that tests should ideally detect even a single cell in a food sample, and this is conventionally achieved by culturing the organism to yield enough cells to allow detection. For *Salmonella* this routinely involves a pre-enrichment in a non-selective medium (to allow stressed and/or injured cells to recover and multiply) followed by a selective enrichment (which encourages multiplication of *Salmonella* but ideally suppresses that of other bacteria). By further sub-culturing the enriched cultures on to solid media, colonies characteristic of *Salmonella* can be 'confirmed' by further biochemical and serological (antibody based) tests. It is the need for successive culturing that makes conventional bacteriological analyses so lengthy - around 4-7 days in the case of *Salmonella*.

The BAX™ PCR system, for example, targets a *Salmonella*-specific DNA sequence and has been successfully applied to a range of meat and dairy products. Adoption of new approaches such as this requires thorough evaluation and validation of the method - both by the manufacturers (as part of method development) and independently. This involves comparing the PCR method with established methods in terms of sensitivity (how few cells the method can detect), the specificity (does it miss any target organisms or falsely detect any non-target organisms?), rapidity, and

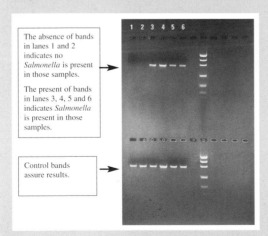

The absence of bands in lanes 1 and 2 indicates no *Salmonella* is present in those samples.

The present of bands in lanes 3, 4, 5 and 6 indicates *Salmonella* is present in those samples.

Control bands assure results.

variability (e.g. between food types, between samples, between operators/ laboratories).

On the basis of this type of evaluation, PCR has progressed from a research tool to an accepted 'off-the-shelf' method; it has been proved to be reliable with an expanding range of foods and to enable detection of as few as 200 cells at much earlier stages in the cultural enrichment process.

References:

1. Bennett, A.R., Greenwood, D., Tennent, C., Banks, J.G. and Betts, R.P. (1996) Use of the BAX™ system, a commercial PCR-based system for the detection of *Salmonella* in foods. CCFRA R&D Report No. 28

2. Bennett, A.R., Greenwood, D., Tennent, C., Banks, J. and Betts, R.P. (1996) Evaluation of the BAX™ system for screening *Salmonella*: a rapid PCR-based system for the detection of foodborne *Salmonella*. CCFRA Research Summary Sheet 1996/1

3. Mroziaski, P.M., Betts, R.P. and Coates, S. (1998) Performance tested method certification of BAX™ for screening/*Salmonella*: case study. Journal of AOAC International, *81*, 1147-1154.

Quantitative PCR

Conventionally, PCR analyses rely on gel electrophoresis to obtain size-estimates of amplified DNA fragments: the PCR product is compared with DNA fragments of standard size on an electrophoretic gel. The use of labelled primers and/or probes, however, allows quantitative determination of the PCR products and, by inference, of the original level of target. In the TaqMan™ system, for example, the PCR involves not only the two primers (which, as is standard, are complementary to either end of the target sequence) but also a probe which recognises the mid-region of the target sequence. This probe carries a reporter dye (fluorescent) and a quencher dye (which quenches the fluorescence). When the two dyes are attached to the probe there is no net fluorescence as the reporter dye is quenched. During PCR, the probe is displaced, leading to cleavage of the reporter dye from the probe. Without an 'adjacent' quencher dye, the reporter signal is no longer quenched, creating a fluorescent signal. The amount of fluorescence emitted can be measured and related back to the level of target DNA.

In contrast, the DARAS™ system uses an enzyme-driven colour reaction as the assay signal. The amplified sequence (target DNA) is captured by a probe immobilised on beads (similar to the format shown in Figure 18). This target can then be detected by a detection probe (in a sandwich configuration) and then by a biotin-avidin-enzyme complex (similar to the indirect sandwich ELISA - see Figure 5). Again, the intensity of colour can be measured, offering the potential to estimate the original level of target material.

Table 14 - Published examples of PCR in food authenticity testing

- Meat species determination
- Detection of GM foods
- Fish and caviar species testing
- Wheat species in pasta
- Meat sex determination
- Detecting bovine material in animal feeds
- Rice authenticity

Box 5 - Case study

Detecting GM foods by PCR

The use of genetic modification in food production had become a high profile issue in Europe, and particularly the UK, by 1999. In addition to legal requirements to label food ingredients containing detectable protein or DNA from GM sources, a number of food companies developed policies aimed at eliminating GM ingredients, to be implemented via quality management systems. Policing the regulations and quality systems required methods for detecting GM materials in foods.

This presented several technical difficulties. First, the GM ingredient might be only a small component of the food - perhaps constituting as little as 1% of the final product - so that any modified DNA or protein may be present in tiny amounts. Second, many food processing operations (e.g. heating, milling) break down the DNA and/or denature proteins. Third, tests to identify specific DNA sequences or proteins require reagents (e.g. probes, primers, antibodies) specific to particular target molecules, which makes 'blanket screening' for all GM materials much more difficult. Finally, it was felt that the ideal method should allow quantitative detection of GM material to help distinguish between large-scale use of a GM ingredient and adventitious low-level contamination.

The first two problems are largely surmounted by using PCR, which is extremely sensitive and which can be used to detect even relatively short sequences of fragmented DNA - though there are many processed materials for which the tests that work on raw materials have yet to be validated. The third problem has been addressed by targeting sequences most commonly introduced during genetic modification. These sequences include the so-called regulatory genes such as 'promoters' (involved in switching on the introduced gene) and terminator regions (which mark the end of the gene). For example, in many genetic modifications, the gene of interest is introduced into the crop as part of a 'genetic construct' which includes a promoter from the cauliflower mosaic virus (CaMV) and a terminator from a bacterial nopaline synthase gene (NOS). By using primers that target the CaMV and NOS sequences in question, therefore, it has been possible to develop a PCR test for application to a range of GM materials. Finally, the problem of quantification is addressed by using quantitative PCR.

For industrial and enforcement purposes, however, the availability of an analytical method is often not in itself enough. To assure maximum reliability of test data, analytical methods should be validated through 'blind' trials and, ideally, the laboratory performing the method should be accredited by a third party and be a participant in a performance proficiency scheme. As individual tests progress from laboratory prototypes to fully validated methods conducted in independent and accredited laboratories, so confidence can grow in the data from analysis of foods for GM ingredients.

References:

1. Boyce, O. Burke, G. and Henehan, G. (1998) Detecting genetically modified foods. Food Science & Technology Today, 12, 213-216

2. Gachett, E., Martin, G.G., Vigneau, F. and Meyer, G. (1999) Detection of genetically modified organisms (GMOs) by PCR: a brief review of methodologies available. Trends in Food Science and Technology, 9, 380-88

3. Garrett, S., Bassett, A. and Brown, H. (1998) Validation study for the screening of GMOs in soya and maize. CCFRA Research Summary Sheet 1998/27

4. Paine, K. and Garrett, S. (1998) The detection of genetically modified DNA in spiked flour samples using the DARAS™ nucleic acid analysis system. CCFRA Research Summary Sheet 1998/30

5. Garrett, S., Paine, K. and Brown, H. (1998) Quantitation of meat species in meat and meat products using Taqman™ PCR. CCFRA Research Summary Sheet 1998/26.

6. Jones, L. (1999) Genetically modified foods. British Medical Journal, 318, 581-584

DNA profiling - RAPD

RAPD stands for randomly amplified polymorphic DNA, though it it sometimes also referred to as AP-PCR (arbitrary primer PCR). It is a form of PCR which allows DNA profiles to be created. The key to RAPD lies in the primers, which are very short (e.g. 10-20 nucleotides long) and have a fixed but arbitrary sequence.

Purely on the basis of probability, with the target genome millions of nucleotides long, sequences which match the sequence of the primer will occur at various points in the genome. Where 2 primers bind on opposite strands and in close proximity (see Figure 21), then the intervening region will be amplified by PCR. The same primers will bind at different sites in different genomes (e.g. in different crop varieties), creating different numbers and sizes of fragments. These are separated on the basis of size by electrophoresis, and the gel stained using a standard fluorescent

Figure 21 - The principle of RAPD
(randomly amplified polymorphic DNA)

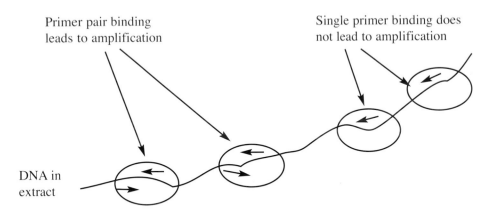

Target DNA is mixed with other PCR ingredients (primers, polymerase and nucleotides) and treated as for PCR. The crucial difference is that the very short random primers will hybridise (recognise and stick to) with many sites in the target DNA. Where two sites are in close proximity and on opposite strands, the PCR will take place, generating fragments. Different length fragments will be obtained if the target DNAs are different. The fragments can be separated by electrophoresis to give a 'banding pattern' on a gel.

DNA stain. Different banding patterns (i.e. DNA profiles or genetic fingerprints) will be obtained for different genetic types.

In practical terms, RAPD involves 2 or 3 main steps - extraction of target DNA from the sample (where necessary), the amplification reaction and electrophoretic analysis of the DNA fragments generated. The whole procedure takes around one working day, involves sophisticated equipment and requires skilled personnel. However, as it is based on amplification, the technique is extremely sensitive. Some concerns have been expressed about reproducibility of the patterns obtained because even minor variations in the method can cause significant changes in the banding patterns.

DNA profiling - AFLP

AFLP stands for amplified fragment length polymorphism. AFLP is basically a combination of RFLP and PCR. The target DNA is digested with a restriction enzyme, just as for RFLP analysis, but is then amplified through PCR (to generate fragments that can be detected on a gel) rather than detected with a probe (as in RFLP analysis). The result is a banding pattern, as with RAPD. The approach contrasts with PCR-RFLP in which the amplification is carried out before the fragments are digested (with a restriction enzyme) to create a characteristic banding pattern. AFLP takes at least one working day and requires specialist skills and equipment, though it is extremely sensitive and potentially very powerful as a laboratory-based method of DNA profiling.

You've bean caught

In Arizona in 1993, a man was convicted of murder on the basis of, amongst other evidence, DNA fingerprints of seed pods of the palo verde tree. Pods found in the man's truck were an exact genetic match with one of the trees found at the murder scene, contradicting his claim that he had never visited the site.

Reference:

1. Anon. (1993) DNA clue. New Scientist 19 June 1993 p12.

Families re-united

The re-uniting of orphaned children with their grandparents in Argentina is a particularly poignant example of use of DNA testing (PCR and RFLP) to confirm familial links. During military rule many babies, whose parents became the victims of genocide, were placed with military families. By the time of the return to democracy in 1983, the true grandparents of the babies had amassed much circumstantial evidence on the whereabouts of many of the, by now, children. They enlisted the help of an American geneticist to confirm grand-parenthood, and with DNA evidence substantiating their claim they were granted custody by the courts (1).

Reference:

1. Knight, P. (1989) DNA analysis reunites Argentinian families. Biotechnology 7 1126.

3.4 DNA methods - advantages and limitations

In general terms, DNA methods offer several distinct advantages over the relevant conventional approach. For example, DNA tests are generally very sensitive so that in some cases multiple tests can, in theory, be carried out on the same small sample. Also, in many cases the tests can be constructed so as to be very specific (e.g. to generate DNA fingerprints specific to individuals) or to be common to particular groups (e.g. variety, strain, species), which provides useful flexibility. Furthermore, because the DNA is the same in all parts of an organism, the same test can be applied to different tissues to obtain the same result - for example, during plant breeding, genetic profiling can be conducted on seed material without having to wait for development of a mature plant. The DNA profile of a living organism is also unaffected by its environmental conditions, which can, however, impact on conventional test methods - for example DNA methods will detect viable and injured microorganisms whereas the latter might be missed by conventional culturing techniques.

Some of these advantages are, however, double edged swords. The sensitivity of the methods can make them more prone to error arising through contamination, so that stringent quality controls have to be employed. As DNA is relatively stable, it cannot, alone, be used to distinguish between viable and dead material - which might be crucial in the context of foodborne pathogens. Finally, although the methods can be intricate and demand sophisticated laboratories, significant progress in method automation and simplification is being made. Overall, DNA methods, when used properly, offer many advantages. The key point is to ensure that the selected method addresses the need of the user and generates an appropriate and reliable end result.

3.5 Recap of the basic principle

DNA diagnostic methods rely on the complementarity between DNA's base pairs, the controlled digestion of DNA with restriction enzymes and/or the biochemical copying of DNA. They also rely on an effective detection system whether this is separation and visualisation of DNA by electrophoresis, a visible reaction on a solid phase or a combination of the two. These ingredients can be brought together in a bewildering array of assay formats. Examples described include the simple dot-blot, RFLP, PCR and its main variants, RAPD and AFLP. The key point, however, is that DNA methods can be extremely powerful and reliable in detecting individual components of complex mixtures - including food components, contaminants (chemical and microbiological) and adulterants.

The case of the Indian rhino

In efforts to protect the Indian rhinoceros (*Rhinoceros unicornis*), scientists in India developed a DNA test to identify rhino horn material and distinguish it from that of other species such as the African double-horned black rhinoceros (*R. bicornis*). The test will allow more effective policing of the illegal dealing in Indian rhino horn material, which is believed by some to possess aphrodisiac properties.

Reference:

Anon. (1998) Probe to catch rhino poacher. Nature Biotechnology 16 991.

Poached eggs

Poaching of live animals is also being attacked through DNA analysis. In one case several hyacinth macaws, which with around only 3000 remaining in the wild are an endangered species, were being sold as the offspring of a captive pair. DNA analysis confirmed non-maternity suggesting that the 'offspring' were actually wild birds. The evidence was used to convict the poachers.

Reference:

1. Kirby, L. (1990) DNA Fingerprinting: An Introduction. Stockton Press. ISBN 0 333 54024 7

4. ATP METHODS

ATP stands for adenosine triphosphate. It is a chemical natural to all living organisms and which consequently is found both in foodborne microbes and food itself. ATP is a high energy compound - an important store of chemical energy. In living cells, various biochemical reactions release energy which is harnessed as ATP. Other reactions, which require and consume energy, are fuelled by the release of energy from ATP. In these living systems, ATP is the energy 'currency' - allowing the chemical transfer of energy from energy-generating to energy-consuming processes.

One of the most visually appealing examples of this is in the firefly *Photinus pyralis* - the fly that flashes light in the dark. It does this for one very good reason: to attract a mate as part of a fascinating courtship ritual - male flies flash around once every five seconds while flying, and females signal back with a flash around two seconds later.

These light flashes are powered by ATP: the fly harnesses the energy released during respiration, in the form of ATP, and then converts this chemical energy into light energy in a relatively simple biochemical reaction. This reaction is known as ATP bioluminescence and has been recreated in the test-tube so that it now forms the basis of a whole range of rapid tests for monitoring the hygienic condition of food contact surfaces.

4.1 Ingredients and assay principles

Apart from the energy input from ATP, there are two main ingredients in the chemical reaction underlying ATP bioluminescence: an enzyme called luciferase and its substrate, called luciferin. During the reaction, the luciferin becomes oxidised to oxyluciferin with the release of light (Figure 22). The amount of light emitted is measured using a luminometer, and expressed in Relative Light Units (RLUs). As one photon of light is emitted for each molecule of ATP consumed, the amount of light emitted is directly related to the amount of ATP present, which in turn can be used as an indication of the level of surface contamination.

Figure 22 - Principle of the ATP-driven bioluminescence reaction

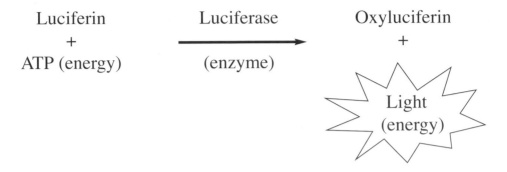

In a reaction fuelled by ATP, the enzyme luciferase catalyses the oxidation of luciferin to oxyluciferin with the release of light energy. For each molecule of ATP consumed, one photon of light is emitted.

Figure 23 - General approach to hygiene testing using ATP bioluminescence

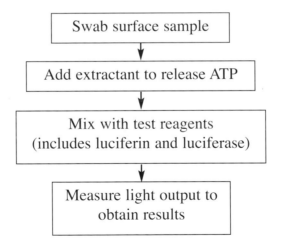

After swabbing a surface, the ATP present in the swab (e.g. from surface microbes or food soil) is released using an extraction buffer, mixed with the other assay reagents (luciferin and luciferase) and incubated under appropriate conditions. The light emitted by the reaction is measured with a luminometer.

4.2 Assay formats

As the ATP bioluminescence reaction is essentially a chemical reaction, it will happen outside of living cells, so long as all the ingredients are present under the right conditions. A range of ATP-based test kits are available for hygiene monitoring, and each provides both the required ingredients and the necessary conditions for the test to work routinely. Although the kits exploit the same basic chemical reaction and overall approach (Figure 23), they vary in how the test sample is captured and prepared, and in the luminometers used to measure the light output.

For example, some of the kits are swab-based. The food contact area of interest is swabbed. The swab is then mixed with ATP-releasing reagents and the reaction ingredients, and then placed directly into the luminometer measurement chamber. In contrast, in cuvette-based systems, the ATP present on the test swab is extracted into a diluent. The diluent is mixed with the reaction ingredients in a cuvette and the light emission measured in a luminometer. The RLU readings from different luminometers available are not directly comparable but those from the same machine are.

4.3 Benefits and limitations of the approach

Rapid tests based on ATP-bioluminescence offer significant advantages as an approach to hygienic monitoring but are also subject to limitations. The primary advantages are their ease-of-use and speed. Their relative simplicity reduces the need for advanced technical skills and laboratory equipment, which widens their accessibility to the food industry.

The rapidity with which results can be obtained (minutes) compared to conventional microbiological analyses (which take several days) brings a number of benefits. For example, it means that they can be used to monitor the effectiveness of cleaning procedures, and to help target further cleaning at areas where this might be necessary (see Box 6). It also allows real-time, focused monitoring of equipment and surfaces at critical control points as part of HACCP (Hazard Analysis and

Box 6 - Case study

ATP tests in hygiene monitoring

In two industrial cleaning trials, the effectiveness of the cleaning regime was assessed by comparing the ATP levels on the food processing surface before and after cleaning. Test samples (swabs) were allocated to Relative Light Unit (RLU) groups - the lower the group number then the lower the ATP level and the cleaner the surface. Preliminary comparisons with conventional microbiological analyses can help establish a cut-off point at which surfaces might be deemed to be have been cleaned sufficiently - say RLU group 2 or below.

In Case A, the pre-cleaning samples fell into the higher RLU groups while the post-cleaning samples fell into the lower RLU groups, demonstrating that cleaning was effective at all sites sampled. In contrast, in Case B, even after cleaning, some surface samples fell into RLU group 4, and so would require re-cleaning under the above criteria.

Case A

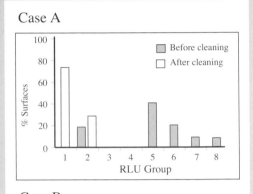

Case B

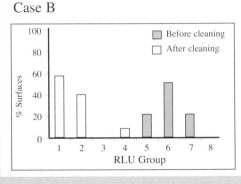

Reference:

1. Holah, J., Gibson, H. and Hawronskyj, J.M. (1995) The use of ATP bioluminescence to monitor surface hygiene. The European Food & Drink Review, Autumn 1995, 82-88

Critical Control Point) making it the only 'microbially based' system. Finally, such rapid results are ideal for training exercises for cleaning personnel, to quickly demonstrate the presence or otherwise of debris.

There are some drawbacks to the tests. As ATP is common to food and microbes alike, a positive ATP result does not necessarily indicate the presence of microbes, and similarly does not allow discrimination between innocuous microbes and food-borne pathogens. Also, expressed as relative light units, the results are in a format which is unfamiliar to some food microbiologists. Finally, in order to implement a system routinely, significant validation work is usually undertaken to assess the appropriateness of the tests for specific processing environments. These might explore factors such as the nature of the surfaces to be assessed, the number of samples needed to generate meaningful data, the effect of cleaning chemical residues on assay performance, and the levels of ATP that occur before and after cleaning.

5. CONCLUSIONS

This guide set out to explain the basic principles underlying modern molecular methods used in food analysis. It covers immunoassays, DNA based tests and ATP methods. It also illustrates the types of analytical problem to which these tests have been applied. These problems, and the associated requirements of the analyst, vary considerably. Some require the detection and identification of single specific components in complex mixtures - be it, for example, the reliable identification of a bacterial pathogen isolated from food or verification that a food conforms to a compositional specification (e.g. suitable for coeliacs). In other instances, the requirement has been to discriminate between closely related strains of the same species (e.g. to trace a source of contamination) or to monitor 'real-time' the effectiveness of a cleaning regime.

Between them, the methods described and illustrated provide the food analyst with extremely powerful tools. Each approach has its own strength (e.g. simplicity, sensitivity, specificity, speed) and each its own limitations (e.g. complexity, highly skilled operators). While the responsibility for choosing the most appropriate tool will fall to the analyst, the background provided here will help those that need to commission analyses, question particular approaches or understand the literature accompanying the latest diagnostic kits, to do so. For those who wish to pursue the topic further, the following section provides references to a range of review and other articles that themselves provide overviews of specific aspects of the technology. These in turn lead to much more detailed and specific articles and, in some cases, contact details for diagnostic kit manufacturers.

REFERENCES AND FURTHER READING

The following titles were used in the preparation of this guide and provide a useful start for reading around the subject. Many of the latest developments reported in the scientific literature can easily be identified through searches of abstract databases.

Archer, D.L. (Ed) The validation of rapid methods in food microbiology. Food Control, **7** (1), 3-58.

Beier, R.C. and Stanker, L.H. (1996) Immunoassays for residue analysis - food safety. American Chemical Society Symposium 621, ISBN 0 8412 3379 9.

Bennett, A.R. and Betts, R.P. (1993) Evaluation of Dynabeads™ anti-*Salmonella* for the isolation of *Salmonella* from herbs and spices. CCFRA Technical Memorandum No. 695.

Bennett, A.R., Greenwood, D., Tennent, C., Banks, J.G. and Betts, R.P. (1996) Use of the BAX™ system, a commercial PCR-based system for the detection of *Salmonella* in foods. CCFRA R&D Report No. 28.

Bennett, A.R., MacPhee, S. and Betts, R.P. (1996) The isolation and detection of *Escherichia coli* O157 by use of immunomagnetic separation and immunoassay procedures. Letters in Applied Microbiology, **22**, 237-243.

Betts, R. (1996) Microbiology the fast way. The European Food & Drink Review, Summer 1996, 45-49.

Betts, R.P. (1997) The catalogue of rapid microbiological methods (3rd edition). CCFRA Review No. 1.

Betts, R. (1998) Foodborne bacteria in the spotlight. Microbiology News, September.

Betts, R., Stringer, M., Banks, J. and Dennis, C. (1995) Molecular methods in food microbiology. Food Australia, **47**, 319-322.

Birstein, V.J., Doukakis, P., Sorkin, B. and De Salle, R. (1997) Population aggregation analysis of three caviar-producing species of sturgeons and implications for the species identification of black caviar. Conservation Biology, 12, 766-775.

Bligh, F. (1999) The development of isotopic analysis and DNA polymorphic markers to determine the geographical and cultivar origin of premium long grain rice. MAFF R&D and Surveillance Report No. 485, MAFF.

Bobbitt, J.A. and Betts, R.P. (1991) Evaluation of the AccuProbe culture confirmation test for Listeria monocytogenes. CCFRA Technical Memorandum No. 630.

Bobbit, J.A. and Betts, R.P. (1992) Confirmation of Listeria monocytogenes using a commercially available nucleic acid probe. Food Microbiology, 9, 311-317.

Boyce, O., Burke, G. and Henehan, G. (1998) Detecting genetically modified foods. Food Science and Technology Today, 12, 213-216.

Broomfield, P.L.E. (1996) Molecular typing of bacterial isolates as a contract service. Microbiology Europe, 4, 24-26.

Carrera, E., Garcia, T., Cespedes, A., Gonzalez, I., Sanz, B., Hernandez, P.E. and Martin, R. (1998) Identification of Atlantic salmon (Salmo salar) and rainbow trout (Oncorhynchus mykiss) by using polymerase chain reaction amplification and restriction analysis of the mitochondrial cytochrome b gene. Journal of Food Protection, 61, 482-486.

Cliver, D.O. (1995) Detection and control of foodborne viruses. Trends in Food Science and Technology, 6, 353-358.

Colquhoun, K.O., Timms, S. and Fricker, C.R. (1998) A simple method for the comparison of commercially available ATP hygiene-monitoring systems. Journal of Food Protection, 61, 499-501.

Dawe, K. and Jones, J.L. (1993) Assessment of immunoassay kits for detection of veterinary drug residues in meat and milk. CCFRA Technical Memorandum No. 691.

Farber, J.M. (1996) An introduction to the hows and whys of molecular typing. Journal of Food Protection, **59**, 1091-1101.

Gachet, E., Martin, G.G., Vigneau, F. and Meyer, G. (1999) Detection of genetically modified organisms (GMOs) by PCR: a brief review of methodologies available. Trends in Food Science & Technology, **9**, 380-388.

Garrett, S., Paine, K. and Brown, H. (1998) Quantitation of meat species in meat and meat products using Taqman™ PCR. CCFRA Research Summary Sheet 26/1998.

Haine, H. and Hall, M. (1997) Rapid methods for detecting ochratoxin A in grain. CCFRA R&D Report No. 52.

Harlow, E. and Lane, D. (1988) Antibodies - a laboratory manual. Cold Spring Harbor Press.

Hawronskyj, J.M. and Holah, J. (1997) ATP: a universal hygiene monitor. Trends in Food Science & Technology, **8** (3), 79-84.

Hilton, A.C., Banks, J.G. and Penn, C.W. (1997) Optimisation of RAPD for fingerprinting *Salmonella*. Letters in Applied Microbiology, **24**, 243-248.

Holah, J., Gibson, H. and Hawronskyj, J.M. (1995) The use of ATP bioluminescence to monitor surface hygiene. European Food and Drink Review, Autumn 1995, pp82-88.

Jones, J.L. (1991) DNA probes: applications in the food industry. Trends in Food Science and Technology, **2**, 28-32.

Jones, K.L., MacPhee, S., Turner, A.J. and Betts, R.P. (1995) An evaluation of the Path Stik for the detection of *Salmonella* from foods. CCFRA R&D Report No. 11.

Jones, K.L., MacPhee, S., Turner, A.J. and Betts, R.P. (1995) An evaluation of the Oxoid *Listeria* rapid test (incorporating Clearview) for the detection of *Listeria* from foods. CCFRA R&D Report No. 19.

Kirby, L.T. (1990) DNA fingerprinting - an introduction. Stockton Press. ISBN 0 333 54024 7

Lantz, P., Hahn-Hagerdal, B. and Radstrom, P. (1994) Sample preparation methods in PCR-based detection of food pathogens. Trends in Food Science and Technology, **5**, 384-389.

MacPhee, S., Tennant, C. and Betts, R.P. (1998) Evaluation of the EiaFoss Campylobacter system for the detection of *Campylobacter* from foods. CCFRA R&D Report No. 58.

MacPhee, S., Bennett, A.R. and Betts, R.P. (1997) Evaluation of the EiaFoss *Listeria* system for the detection of *Listeria* species from food. CCFRA R&D Report No. 45.

Masso, R, and Oliva, J. (1997) Technical evaluation of an automated analyser for the detection of *Salmonella enterica* in fresh meat products. Food Control, **8**, 99-103.

Meng, J., Doyle, M.P., Zhao, T. and Zhao, S. (1996) Detection and control of *Escherichia coli* O157:H7 in foods. Trends in Food Science and Technology, **5**, 179-184.

Meyer, R., Hofelen, C., Luthy, J. and Candrian, U. (1995) Polymerase chain reaction-restriction length polymorphism analysis: a simple method for species identification in food. Journal of AOAC International, **78**, 1542-1551.

Miyamoto, T., Tian, H.Z., Okabe, T., Trevanich, S., Asoh, K., Tomoda, S., Honjoh, K. and Hatano, S. (1998) Application of random amplified polymorphic DNA analysis for detection of *Salmonella* spp. in foods. Journal of Food Protection, **61**, 785-791.

Morgan, M.R.A., Smith, C.J. and Williams, P.A. (Eds) (1992) Food Safety and Quality Assurance - Applications of Immunoassay Systems. Elsevier Applied Science Publications.

Murtiningsih and Cox, J.M. (1997) Evaluation of the Serobact™ and Microbact™ systems for the detection and identification of *Listeria* spp. Food Control, **8**, 205-210.

Nelson, J.O., Karu, A.E. and Wong, R.B. (eds) (1995) Immunoanalysis of agrochemicals - emerging technologies. American Chemical Society Symposium Series 586. ISBN 0 8412 3149 4.

Oscar, T.P. (1998) Identification and characterisation of *Salmonella* isolates by automated ribotyping. Journal of Food Protection, **61**, 519-524.

Querol, A. and Ramon, D. (1996) The application of molecular techniques in wine microbiology. Trends in Food Science and Technology, **7**, 73-78.

Saunders, G.C. and Parkes, H.C. (1999) Analytical molecular biology - quality and validation. Royal Society of Chemistry. ISBN 0 85404 472 8

Stringer, S.C., Dodd, C.E.R., Morgan, M.R.A. and Waites, W.M. (1995) Detection of micro-organisms *in situ* in solid food. Trends in Food Science and Technology, 6, 370-374.

Syernesjo, A. and Johnsson, G. (1998) A novel rapid enzyme immunoassay (Fluorophos BetaScreen) for detection of beta-lactam residues in ex-farm raw milk. Journal of Food Protection, **61**, 808-811.

Tartaglia, M., Saulic, E., Pestalozza, S., Morelli, L., Antonucci, G. and Battaglia, P.A. (1998) Detection of bovine mitochondrial DNA in ruminant feeds: a molecular approach to test for the presence of bovine derived materials. Journal of Food Protection, **61**, 513-518.

Taylor, W.J., Patel, N.P. and Jones, J.L. (1994) Antibody based methods for assessing seafood authenticity. Food and Agricultural Immunology, **6**, 305-314.

Wiseman, G. (1999) Quantitative PCR detection of *T. aestivum* adulteration in commercial *T. durum* pasta using PSR 128 primers: optimisation. MAFF R&D and Surveillance Report No. 486, MAFF.

Zeng, S.S., Hart, S., Escobar, E.N. and Tesfai, K. (1998) Validation of antibiotic residue tests for dairy goats. Journal of Food Protection, **61**, 344-349.

About CCFRA

The Campden & Chorleywood Food Research Association (CCFRA) is the largest independent membership-based food and drink research centre in the world. It provides wide-ranging scientific, technical and information services to companies right across the food production chain - from growers and producers, through processors and manufacturers to retailers and caterers. In addition to its 1500 members (drawn from over 50 different countries), CCFRA serves non-member companies, industrial consortia, UK government departments, levy boards and the European Union.

The services provided range from field trials of crop varieties and evaluation of raw materials through product and process development to consumer and market research. There is significant emphasis on food safety (e.g. through HACCP), hygiene and prevention of contamination, food analysis (chemical, microbiological and sensory), factory and laboratory auditing, training and information provision. As part of the latter, CCFRA publishes a wide range of research reports, good manufacturing practice guides, reviews, videos, databases, software packages and alerting bulletins. These activities are under-pinned by fully-equipped modern food processing halls, product development facilities, extensive laboratories, a purpose-built training centre and a centralised information service.

In 1998 CCFRA established a wholly owned subsidiary in Hungary from where an experienced team of scientists and technologists provides training and consultancy on HACCP, quality management, product development, market and consumer research, food and environment law, and hygiene to Eastern Europe.

To find out more, visit the CCFRA website at www.campden.co.uk